Lessons Learned from Early Implementation of The Maintenance Rule at Nine Nuclear Power Plants

I0468032

Manuscript Completed: June 1995
Date Published: June 1995

C.D. Petrone, R.P. Correia, S.C. Black

Division of Technical Support
Office of Nuclear Reactor Regulation
U. S. Nuclear Regulatory Commission
Washington, DC 20555–0001

NRC FORM 335
(2-89)
NRCM 1102,
3201, 3202

U.S. NUCLEAR REGULATORY COMMISSION

BIBLIOGRAPHIC DATA SHEET

(See instructions on the reverse)

1. REPORT NUMBER
(Assigned by NRC, Add Vol., Supp., Rev., and Addendum Numbers, if any.)

NUREG-1526

2. TITLE AND SUBTITLE

Lessons Learned From Early Implementation of The Maintenance Rule at Nine Nuclear Power Plants

3. DATE REPORT PUBLISHED

MONTH	YEAR
June	1995

4. FIN OR GRANT NUMBER

5. AUTHOR(S)

C.D. Petrone, R.P. Correia, S.C. Black

6. TYPE OF REPORT

Final

7. PERIOD COVERED *(Inclusive Dates)*

N/A

8. PERFORMING ORGANIZATION – NAME AND ADDRESS *(If NRC, provide Division, Office or Region, U.S. Nuclear Regulatory Commission, and mailing address; if contractor, provide name and mailing address.)*

Division of Technical Support
Office of Nuclear Reactor Regulation
U.S. Nuclear Regulatory Commission
Washington, DC 20555-0001

9. SPONSORING ORGANIZATION – NAME AND ADDRESS *(If NRC, type "Same as above"; if contractor, provide NRC Division, Office or Region, U.S. Nuclear Regulatory Commission, and mailing address.)*

SAME AS ABOVE

10. SUPPLEMENTARY NOTES

11. ABSTRACT *(200 words or less)*

This report summarizes the lessons learned from the nine pilot site visits that were performed to review early implementation of the maintenance rule using the draft NRC Maintenance Inspection Procedure. Licensees followed NUMARC 93-01, "Industry Guideline for Monitoring the Effectiveness of Maintenance at Nuclear Power Plants." In general, the licensees were thorough in determining which structures, systems, and components (SSCs) were within the scope of the maintenance rule at each site. The use of an expert panel was an appropriate and practical method of determining which SSCs are risk significant. When setting goals, all licensees considered safety but many licensees did not consider operating experience throughout the industry. Although required to do so, licensees were not monitoring at the system or train level the performance or condition for some systems used in standby service but not significant to risk. Most licensees had not established adequate monitoring of structures under the rule. Licensees established reasonable plans for doing periodic evaluations, balancing unavailability and reliability, and assessing the effect of taking equipment out of service for maintenance. However, these plans were not evaluated because they had not been fully implemented at the time of the site visits.

12. KEY WORDS/DESCRIPTORS *(List words or phrases that will assist researchers in locating the report.)*

Maintenance Rule
Maintenance Rule Pilot Site Visits
10 CFR 50.65

13. AVAILABILITY STATEMENT

Unlimited

14. SECURITY CLASSIFICATION

(This Page)

Unclassified

(This Report)

Unclassified

15. NUMBER OF PAGES

16. PRICE

ABSTRACT

This report summarizes the lessons learned from the nine pilot site visits that were performed to review early implementation of the maintenance rule using the draft NRC Maintenance Inspection Procedure. Licensees followed NUMARC 93-01, "Industry Guideline for Monitoring the Effectiveness of Maintenance at Nuclear Power Plants." In general, the licensees were thorough in determining which structures, systems, and components (SSCs) were within the scope of the maintenance rule at each site. The use of an expert panel was an appropriate and practical method of determining which SSCs are risk significant. When setting goals, all licensees considered safety but many licensees did not consider operating experience throughout the industry. Although required to do so, licensees were not monitoring at the system or train level the performance or condition for some systems used in standby service but not significant to risk. Most licensees had not established adequate monitoring of structures under the rule. Licensees established reasonable plans for doing periodic evaluations, balancing unavailability and reliability, and assessing the effect of taking equipment out of service for maintenance. However, these plans were not evaluated because they had not been fully implemented at the time of the site visits.

CONTENTS

EXECUTIVE SUMMARY

The staff learned several lessons while visiting the nine pilot sites to review early implementation of the maintenance rule (the rule). These sites were visited to verify usability and adequacy of the draft NRC Maintenance Rule Inspection Procedure and determine the strengths and weaknesses with the implementation of the rule at each site. This report presents the results of these site visits to other licensees for their consideration during their implementation of the rule. The major findings for each subject area are summarized below.

Use of Industry Guideline

Licensees implemented the rule using the guidance in NUMARC 93-01, "Industry Guideline for Monitoring the Effectiveness of Maintenance at Nuclear Plants," May 1993, which the NRC endorsed in Regulatory Guide 1.160, "Monitoring the Effectiveness of Maintenance at Nuclear Power Plants," Revision 1, January 1995.

Scoping

Most licensees were thorough in determining which structures, systems, and components (SSCs) are within the scope of the rule at their sites. The licensees correctly classified most safety-related SSCs and those non–safety-related SSCs whose failure could prevent safety-related SSCs from fulfilling their safety-related function. Certain licensees incorrectly missed classifying a few non–safety-related SSCs as being within the rule. These systems are relied upon to mitigate accidents or transients or are used in emergency operating procedures, or their failure could cause a reactor scram or actuation of a safety-related system. These systems included control room annunciators, circulating water, reactor coolant pump vibration monitoring, extraction steam, condenser air removal, screen wash water, generator gas, and turbine lube oil.

Risk Determination

The team found that each licensee had used a well qualified expert panel to determine which SSCs were risk significant. The use of an expert panel that considers probabilistic risk assessment (PRA) insights is an appropriate and practical method of determining risk significance. The expert panels at most sites considered PRA insights using risk reduction worth, risk achievement worth, core damage frequency contribution, and Fussell-Vesely (F/V) importance measures. However, the expert panels at two sites did not receive the results of all importance measures to consider when determining risk significance.

Categorizing Structures, Systems, and Components in Paragraph (a)(1) or (a)(2)

The process and procedures used by most licensees for categorizing SSCs under paragraph (a)(1) or (a)(2) of the rule was reasonable. However, some licensees were reluctant to place SSCs in the paragraph (a)(1) category because having SSCs in that category would imply their preventive maintenance programs were ineffective.

Corrective Actions

Licensees established effective corrective action processes or programs. To implement the maintenance rule, most licensees had assigned primary responsibility for establishing corrective action to the system engineers; a few licensees had assigned that responsibility to an expert panel. Both approaches produced acceptable results.

Safety Consideration in Goal Setting

Licensees established satisfactory programs for taking safety into consideration when setting goals.

Industry Operating Experience

Many licensees' procedures did not have adequate guidance for ensuring that operating experience was considered, where practical, when establishing goals. In addition, licensees had not established a systematic and consistent method of collecting and using SSC reliability and availability data from other licensees when setting goals.

Monitoring and Trending of Systems and Components

The SSC performance or condition trending that was being performed by most licensees was not well coordinated and integrated with the goals and performance criteria established under the rule.

Monitoring and Trending of Structures

Most licensees assigned low priority to the monitoring of structures under the rule. Several incorrectly assumed that many of their structures are inherently reliable. The performance criteria for monitoring some structures were not predictive and did not give early warning of degradation.

Periodic Evaluations

Two licensees had performed a periodic evaluation before the site visit. The team reviewed the results of these evaluations and found that they generally met the requirements of the rule. At the seven other sites, the team reviewed the licensees' preliminary plans and procedures for performing the periodic evaluation and found them to be reasonable.

Balancing Unavailability and Reliability

The preliminary plans established by all nine licensees for balancing unavailability and reliability were reasonable. However, the team was unable to fully evaluate these activities because these plans had not been implemented at the time of the site visits.

Plant Safety Assessments Before Taking Equipment Out of Service

Many licensees developed matrices to define which system combinations could be allowed out of service at the same time. Several licensees are planning to use real time (or *near-real-time*) risk monitors to calculate the risk changes associated with the planned maintenance activities. Both the matrix approach and the risk monitor approach are reasonable ways of assessing the effect on plant safety when taking equipment out of service for monitoring or preventive maintenance. However, the team did not evaluate the effectiveness of either method because neither had been fully implemented.

Conclusion

Upon considering the observations made during these pilot site visits, the team concluded that the requirements of 10 CFR 50.65, "Requirements for Monitoring the Effectiveness of Maintenance at Nuclear Power Plants," can be met by using NUMARC 93-01, "Industry Guidance for Monitoring the Effectiveness of Maintenance at Nuclear Power Plants," if the recommendations in this report are taken into consideration. The team also concluded that the performance-based approach to implementing the rule is practical, the draft inspection procedure can be used to monitor the implementation of the rule, and the existing PRA tools and models, used in conjunction with an expert panel, are adequate for purposes of taking risk into consideration when implementing the rule.

ACKNOWLEDGEMENTS

The authors wish to acknowledge Millard L. Wohl for reviewing risk determination methods and online maintenance evaluations, Francis X. Talbot for evaluating plant systems for inclusion in the scope of the maintenance rule and preparing the tables. The authors also wish to acknowledge Jeffrey Main, who edited this report.

NRC Team Members for the Nine Pilot Site Visits

Thomas Foley, NRR

Ronald Frahm Jr., NRR

Wayne Scott, NRR

Francis Talbot, NRR

Dave Nelson, NRR

Melvin Shannon, NRR

Jin Chung, NRR

Ed Ford, SRI, Waterford

Angel Coello, Spanish Nuclear Safety Council

Steven Barr, Region I

James Stewart, Region I

Herb Williams, Region I

Tom Kenny, Region I

Tom Shedlosky, Region I

Paul J. Kellogg, Region II

Al Walker, Region III

Wayne Shafer, Region III

Clark Vanderneit, Region III

George Replogle, Region III

John Whittemore, Region IV

Licensee Representatives

Roger Murgatroyd, Florida Power Corporation

Mark I. Forsyth, Houston Power and Light

Mark Ajluni, Georgia Power Corporation

Lewis A. Ward, Georgia Power Corporation

Keith A. Fry, Georgia Power Corporation

Thomas Bardauskas, Commonwealth Edison

Jon Anderson, Boston Edison Company

Don Hanley, Boston Edison Company

Jerry A, Kleam, Boston Edison Company

Ron J, Zula, Carolina Power and Light

Martin Bridges, Carolina Power and Light

Robert T. Biggerstaff, Carolina Power and Light

Robert Bickford, Maine Yankee Atomic Power Company

William Drake, Maine Yankee Atomic Power Company

Sammy Mooney, Entergy Operations Inc.

Tom H. Thurmon, Entergy Operations Inc.

NEI Representatives

Dan J Rains

Adrian Heymer

Ray Ng

Tony Pietrangelo

1 INTRODUCTION

1.1 Objective

This report summarizes the lessons learned during the pilot site visits that a team from the U. S. Nuclear Regulatory Commission (NRC) made from September 1994 to March 1995 to review early implementation of the maintenance rule. The maintenance rule (the rule), which was published on July 10, 1991, as 10 CFR 50.65, "Requirements for Monitoring the Effectiveness of Maintenance at Nuclear Power Plants," will take effect on July 10, 1996. These reviews were performed at nine nuclear power plant sites that had volunteered for early review of their programs for implementing the rule. The purpose of these site visits was to verify usability and adequacy of the draft NRC Maintenance Rule Inspection Procedure and to determine the strengths and weaknesses of the implementation of the rule at each site. Other licensees should consider this information when developing programs for implementing the rule. NRC inspectors may also use it while evaluating licensees' implementation of the rule.

1.2 Need for the Maintenance Rule

In the statements of consideration for the rule, the Commission stated that such a rule is needed because proper maintenance is essential to plant safety and because effective maintenance is clearly linked to safety. Good maintenance helps limit the number of transients and challenges to safety systems by ensuring operability, availability, and reliability of safety equipment. Good maintenance is important in ensuring that failures of structures, systems, and components (SSCs) other than safety-related SSCs that could initiate or adversely affect a transient or accident are minimized. Minimizing challenges to

safety systems is consistent with the Commission's defense-in-depth philosophy. Maintenance is also important to ensure that design assumptions and margins in the original design basis are maintained or at least not unacceptably degraded. Therefore, maintenance at nuclear power plants is clearly important in protecting the public health and safety.

The results of the Commission's Maintenance Team Inspections (MTIs) indicated that licensees have adequate maintenance programs and were improving implementation of their programs. However, some common maintenance-related weaknesses were found, such as inadequate root cause analysis leading to repetitive failures, lack of equipment performance trending, and the need for consideration of plant risk in the prioritization, planning, and scheduling of maintenance.

The Commission believes that the effectiveness of maintenance must be assessed continually to verify that key structures, systems, and components are capable of performing their intended function. Further, licensees need to consider revising programmatic requirements where poor assessment results indicate ineffective maintenance.

However, despite significant industry accomplishment in the areas of maintenance program content and implementation, plant events caused by the degradation or failure of plant equipment continue to occur as a result of instances of ineffective maintenance. Additionally, operational events have been exacerbated by, or resulted from, plant equipment being unavailable because of maintenance activities. Most existing requirements and industry maintenance initiatives do not call for licensees to routinely assess the availability of safety significant structures, systems, and components. These events and circumstances attest to the need to continually assess the results of maintenance effectiveness since, together with equipment reliability, equipment availability is an important measure of maintenance effectiveness.

The Commission also recognizes its need to broaden its capability to take timely enforcement action where maintenance activities fail to give reasonable assurance that safety-significant SSCs are capable of performing their intended function. Additionally, the Commission concluded that it is necessary for NRC to include requirements for corrective action to address instances of ineffective maintenance, and for licensees to use the results of monitoring and assessment to improve maintenance programs.

The Commission also had several other reasons for finding the need for a rule requiring that the effectiveness of maintenance be monitored. One of these reasons is that the Commission's current regulations, regulatory guidance, and licensing practice do not clearly define the Commission's expectations for ensuring the continued effectiveness of maintenance programs at nuclear power plants. Another is that industry has no guidance regarding the monitoring of maintenance effectiveness.

Requirements and guidance for monitoring maintenance effectiveness and for taking corrective action when maintenance is ineffective should enhance the Commission's capability to take timely and effective action against licensees with inadequate or poorly conducted maintenance to ensure prompt resumption of effective maintenance activities.

On July 10, 1991, the Commission published the final rule, 10 CFR 50.65, in the *Federal Register*. When the rule takes effect on July 10, 1996, it will require all nuclear power plant licensees to monitor the effectiveness of main-

tenance activities. The rule provides for continued emphasis of the defense-in-depth principle by including select balance-of-plant (BOP) SSCs, integrates risk consideration into the maintenance process, establishes an enhanced regulatory basis for inspection and enforcement of BOP maintenance-related issues, and gives a strengthened regulatory basis for ensuring that the progress achieved is sustained in the future.

1.3 Process-Oriented and Results-Oriented Rulemaking

Although they do not appear to have a formal definition, the terms *process-oriented* (or programmatic, or prescriptive) and *results-oriented* (or results-based, or performance-based) are increasingly being used to describe various rulemaking activities. A process-oriented rule, the traditional approach for most rulemaking, includes detailed requirements or instructions. The advantage to such rules is that they are easier to enforce because the requirements for implementing the rule are delineated in greater detail than would be the case for results-oriented rules. Using a process-oriented rule, licensees generally have a clearer idea of what they need to do to implement the requirements of the rule, and NRC inspectors have a clearer idea of what to inspect. The disadvantage to such rules is that they tend to be inflexible and thus may prevent licensees from using the most efficient and effective means of implementing the rule. Two examples of process-oriented rules are Appendices J (Primary Containment Leakage Testing) and R (Fire Protection Program for Nuclear Power Facilities) to 10 CFR Part 50.[1]

---

[1] Although Appendix J rule is a good example of a process-oriented rule, the NRC is revising this rule to add a results-oriented option permitting licensees the flexibility to adjust leak rate test frequencies based on performance.

These rules contain detailed requirements for test frequency, test pressures, training, and record keeping.

A results-oriented (or results-based or performance-based) rule describes, in general terms, the results expected while leaving the details of achieving those results up to the licensee. Such a rule has the advantage of allowing the licensee to devise the most effective and efficient means of achieving the results described in the rule. It also allows licensee to consider safety or risk significance when developing its programs. The disadvantage of a results-oriented rule is that it may be difficult to enforce because the requirements for compliance may be less clearly defined than the requirements of a process-oriented rule. The maintenance rule, 10 CFR 50.65, is a results-oriented rule.[2] Although licensees clearly prefer results-oriented regulations over process-oriented regulations, such regulations lack of detail as became apparent during the development of regulatory guidance for the rule. The Nuclear Management and Resources Council, Inc, (NUMARC) now the Nuclear Energy Institute (NEI), representing the industry, asked that the inspection procedure (which is normally developed after the regulatory guidance has been issued) be prepared early and given them to use while preparing the industry guidance document. They wanted to use the inspection procedure to address details, not in the rule itself. Results-oriented rules place a greater

---

[2] Although the maintenance rule has been described as a results-oriented rule, it prescribes certain specific programmatic actions. For example, paragraph (a)(3) requires that a periodic evaluation be performed and that the effect on total plant risk be considered before removing equipment from service for maintenance. Therefore, while the maintenance rule is much more results-oriented and less programmatic than most other existing rules, it has certain process-oriented elements.

burden on licensees to develop the supporting details needed to implement the rule.

1.4 Description of the Maintenance Rule

Section (a) of the rule contains most of its detailed technical requirements; paragraph (b) defines the scope of structures, systems, and components within the rule; and paragraph (c) states that the licensees shall implement the rule no later than July 10, 1996. Section (a) consists of paragraphs (1), (2), and (3).

1.4.1 Goals and Monitoring

Paragraph (a)(1) of the rule requires the operator of each power reactor to set goals and to monitor the performance or condition of SSCs in a manner sufficient to give reasonable assurance that those SSCs are capable of performing their intended functions. The rule states the goals must be commensurate with safety and where practical take into account industry-wide operating experience. The rule also requires operators to take appropriate corrective action when the performance or condition of an SSC does not meet established goals. In keeping with the non-prescriptive intent of the rule, the licensee establishes the goals, not the NRC.

1.4.2 Effective Preventive Maintenance

Paragraph (a)(2) of the rule establishes an alternative approach to the monitoring regime required by paragraph (a)(1) of the rule. The NRC recognizes in this approach that, in certain cases, the performance or condition of SSCs could be effectively controlled by doing adequate preventive maintenance rather than by monitoring against goals.

1.4.3 Periodic evaluation and safety assessments

Paragraph (a)(3) of the rule requires that performance- and condition-monitoring activities and associated goals and preventive maintenance activities be evaluated at least every refueling cycle, provided the interval between evaluations does not exceed 24 months. This paragraph will require licensees to systematically review activities under paragraphs (a)(1) and (a)(2) of the rule and to adjust those activities where needed. These evaluations are required by the rule to take into account, where practical, industry-wide operating experience.

Paragraph (a)(3) of the rule also requires that adjustments be made, where necessary, to ensure that the objective of preventing failures of SSCs through maintenance is appropriately balanced against the objective of minimizing the time SSCs are unavailable because of monitoring or preventive maintenance. This requirement recognizes that performing monitoring or preventive maintenance often requires that the SSCs be taken out of service, rendering them unavailable for operation. The higher reliability gained by increased monitoring or preventive maintenance could decrease availability, and possibly impair safety.

Paragraph (a)(3) of the rule also states that in performing monitoring and preventive maintenance activities, an assessment of the total plant equipment that is out of service should be considered to determine the overall effect on performance of safety functions. To address this element of the rule, licensees should continually evaluate whether voluntary removal of equipment from service to perform monitoring and preventive maintenance activities may place the plant in a less safe condition, especially if other supportive equipment is out of service. An example of this type of situation might be taking one train of a safety system out of service while one of the alternate sources of power supply for the redundant train is also out of service. Although the technical specification operability requirements address this concern, the NRC staff recently found vulnerabilities not addressed by the technical specifications. The requirements of this paragraph of the rule extend beyond those of the technical specifications.

1.4.4 Scope

Paragraph (b) of the rule defines those SSCs that must be included within the scope of the rule. They include all safety-related SSCs and those non–safety-related SSCs (1) that are relied upon to mitigate accidents or transients or are used in emergency operating procedures (EOPs), (2) whose failure could prevent safety-related SSCs from fulfilling their intended functions, or (3) whose failure could cause a scram or safety system actuation.

1.5 Implementation

To develop implementation guidance, the NRC and NUMARC established parallel steering and working groups. In June 1993, the NRC published Regulatory Guide 1.160, "Monitoring the Effectiveness of Maintenance at Nuclear Power Plants," which endorsed NUMARC 93-01, "Industry Guideline for Monitoring the Effectiveness of Maintenance at Nuclear Power Plants," May 1993. NUMARC sponsored two industry workshops in August 1993 to educate the industry on the methods stated in NUMARC 93-01 for implementing the rule. The NRC staff participated in these workshops.

The NRC staff developed a draft inspection procedure to verify implementation of the rule. On March 31, 1994, NRC sponsored a public workshop in Rockville, Maryland, where members of the public and the nuclear industry could ask questions about the inspection procedure. At the workshop, NRC explained its expectations about implementation of the rule.

Beginning in September 1994 and ending in March 1995, the NRC staff visited nine pilot sites to validate the draft inspection procedure. The licensees for these plants had voluntarily implemented most of the requirements of the rule, which does not become effective until July 10, 1996. NRC coordinated with NEI in selecting the sites: Grand Gulf, Maine Yankee, Shearon Harris, Pilgrim, Byron, Hatch, Vogtle, South Texas, and Crystal River. The NRC review teams included representatives from the Quality Assurance and Maintenance Branch and the Probabilistic Safety Assessment Branch from the Office of Nuclear Reactor Regulation (NRR), representatives from the Trends and Patterns Analysis Branch from the Office for Analysis and Evaluation of Operational Data (AEOD), and regional inspectors. This report summarizes the lessons learned from the site visits for the benefit of all licensees and NRC inspectors and for the staff to use in completing the inspection guidance.

2 LESSONS LEARNED FROM PILOT SITE VISITS

The NRC staff documented their findings, conclusions, and recommendations in letters to the individual licensees following each site visit. These reports addressed each of each requirement of the rule, such as scoping and risk determination, as a separate topic.

2.1 Use of Industry Guideline NUMARC 93-01

The guidance in NUMARC 93-01 was endorsed in Regulatory Guide 1.160 as containing acceptable methods of implementing the rule. All nine licensees used NUMARC 93-01 for implementing the rule. Eight licensees took minor exceptions that the NRC review team reviewed and found acceptable. One licensee took major exceptions that the team reviewed and also found acceptable.

2.2 Scoping

Before visiting each site, the team reviewed the plant final safety analysis report (FSAR), EOPs, and Individual Plant Evaluation (IPE) insights to select sample SSCs that the team believed were within the scope of the rule. The team used this sample list to determine whether the licensee had classified the required SSCs as being within the scope as defined in paragraph (b) of the rule.

At each site, the team reviewed the process and procedures used by the licensee to determine which SSCs were to be included from the scope of the rule and which were excluded.

Findings

Most licensees followed the scoping process described in NUMARC 93-01 and included the majority of SSCs the team believed should be within scope. However, the following are examples of excluded SSCs from various sites that the team believed should be within the scope of the rule at that particular facility:

1. **Control room annunciators.** The justification for excluding control room annunciators, which constitute a non-safety-related system used in EOPs, from the scope of the rule did not follow the guidance in paragraph 8.2.1.3 of NUMARC 93-01, which allows SSCs covered by paragraph 50.65(b)(2)(i) to be excluded if they do not add "significant value to the mitigation function." The licensee based its decision on the fact that the annunciators are redundant to other safety-related instruments such as gauges and chart recorders, which are considered the primary instruments used in EOPs. However, the annunciators add significant value to the mitigation function because they will often give the first warning of an out-of-tolerance condition or developing accident. The team discussed this concern with several of the licensee's control room operators, who confirmed that control room annunciators often gave them the first warning of a problem. The team concluded that those annunciators, which are used in the EOPs, do add significant value to the mitigation function and therefore are within the scope of the rule.

2. **Site grounding system.** The licensee's documentation states that the site grounding system is a "non–safety-related SSC whose failure causes trip/power reduction." Therefore, the site grounding system should be included within the scope of paragraph 50.65(b)(2)(iii) of the rule. The licensee excluded this system from the

scope of the rule because "the plant has not experienced significant problems with system in the past." The team believes that the licensee has misinterpreted paragraph (b)(2)(iii) of the rule, which states that non–safety-related SSCs "whose failure *could cause* a reactor scram or actuation of a safety-related system" [emphasis added] shall be included within the scope of the rule. The licensee has interpreted this requirement to mean *has caused* rather than *could cause*. The team believes that this interpretation is incorrect.

3. **Circulating water system**. Paragraph (b)(2)(iii) of the rule requires that all non–safety-related SSCs whose failure *could cause* a reactor scram or the actuation of a safety-related system be included within the scope of the rule. The team believes that the circulating water system falls under this provision since its failure could cause a turbine trip and subsequent reactor scram. The licensee's representative stated that they had considered including the circulating water system under the scope but had decided against it for several reasons. The team members discussed each of these reasons with the licensee's representatives and stated the team's views on each of these reasons. A summary of the issues follows.

High reliability. The licensee stated that the design of the circulating water system makes it very reliable. The cooling pond is man-made and is therefore not subject to the causes of fouling (debris, plant, or animal life) that plague natural sources of cooling water at other reactor sites. They stated that reliability this is evidenced by the fact that the site has never experienced a loss of circulating water that resulted in a reactor scram or safety system actuation. Therefore, they concluded that the

circulating water system did not meet the criteria of paragraph (b)(2)(iii) of the rule.

The team disagreed with this position. Paragraph (b)(2)(iii) of the rule states that non–safety-related SSCs whose failure could cause a reactor scram or safety system actuation should be included within the scope of the rule. Nothing in the rule itself, the statements of consideration, Regulatory Guide 1.160, or NUMARC 93-01, indicates that the reliability should be considered when determining whether a system is under the scope of the rule. The phrase beginnning with *whose failure* means that the system is assumed to fail. The only question is whether or not the assumed failure could cause a subsequent reactor scram or safety system actuation.[3]

Redundant trains The licensee also made the argument that since the circulating water system has four trains (each consisting of screens, pumps, and piping) it would be unreasonable to assume that the whole system could fail. If only one train failed, the plant could continue operation, or take action to reduce power and avoid a reactor scram or safety system actuation.

The team disagreed with this position. A failure of the system must be assumed since it would be the worst case scenario.

---

[3] Paragraph (b)(2)(ii) of the rule requires that non–safety-related structures and components "whose failure could prevent safety-related structures, systems, and components from fulfilling their safety-related function" be included within the scope of the rule. Similar to the argument for paragraph (b)(2)(iii), the words *whose failure* mean that the failure is assumed. The criteria are only intended to determine if the assumed failure of the system could prevent a safety-related system from fulfilling its safety-related function. Therefore, the system is assigned a reliability of zero for this analysis because it is assumed to fail.

The licensee's approach of assuming a train failure is not conservative and does not meet the intent of paragraph (b)(2)(iii).

Operator action The licensee made the argument that in the unlikely event that the entire circulating water system failed, the reactor operators would be able to recognize the event in sufficient time to take mitigating actions that would prevent a reactor scram or safety system actuation.

The team disagreed with this position. Operator action, or the lack of it, is not a criterion to be considered when making a scoping determination under 10 CFR 50.65(b). Operator actions are not discussed as a criterion in either the rule, the statements of consideration, Regulatory Guide 1.160, or NUMARC 93-01.

Directly cause. Paragraph (b)(2)(iii) of the rule requires that the scope of the rule include all non–safety-related SSCs whose failure *could cause* a reactor scram or the actuation of a safety-related system. The licensee interpreted this criterion to mean that only those SSCs that *directly* cause a reactor scram should be considered. Therefore, a loss of circulating water that caused a turbine trip and subsequently resulted in a reactor scram should not come under the scope of the rule because it would not be a direct trip.

The team disagreed that *could cause* should be interpreted to mean *directly cause*. Nothing in the rule itself, the statements of consideration, Regulatory Guide 1.160, or NUMARC 93-01, indicates that only those non–safety-related SSCs that *directly* cause a reactor scram or safety system actuation should be included within the scope of the rule. The licensee's use of *directly cause* is not a conservative interpretation of the rule.

Hypothetical failures. The licensee stated that a failure of the circulating water system was a hypothetical failure because it had not happened at their site and was not described in the safety analysis report or any other analysis. Therefore, they believed that they were not required to consider such a failure under the guidance in paragraph 8.2.1.5 of NUMARC 93-01.

The team disagreed with the position taken by the licensee. Paragraph 8.2.1.5 of NUMARC 93-01 states that "[t]he determination of hypothetical failures that could result from system interdependencies but have not been previously experienced is not required." In the section "High Reliability" of NUMARC 93-01, the authors state the fact that the circulating water system has not failed in the past does not make this a hypothetical failure. If the circulating water system does fail, the interactions that could result in a reactor scram are not hypothetical; rather they are the normal expected sequence of operation of the plant. A loss or reduction in circulating water would require a turbine runback or trip, which could result in a reactor scram. The team believes that the *hypothetical failures* described in NUMARC 93-01 refer to two or more events (not previously experienced or analyzed) that occur simultaneously and result in a reactor scram or safety actuation.[4]

Other examples. Other examples of SSCs that the team believed should have been within the scope of the rule at particular sites follow: shield walls that separate the

---

[4] Some of these issues were discussed with other licensees during previous site visits. This discussion is additional guidance to all licensees. The team will recommend revising the implementing guidance to further clarify this issue.

station startup transformers, plant computer, heat tracing and freeze protection, reactor coolant pump vibration monitoring, lightning protection, cathodic protection systems, extraction steam, condenser air removal, screen wash water, gland steam, gland seal water, generator gas, turbine lube oil, and turbine generator seal oil.

One licensee used a different interpretation of the guidance contained in NUMARC 93-01 for determining which non–safety-related SSCs are relied upon to mitigate accidents or transients or are used in plant EOPs under paragraph (b)(2)(i) of the rule. NUMARC 93-01, paragraph 8.2.1.3, states that only those SSCs that "add significant value to the mitigation function of an EOP by providing the total or a significant fraction of the total functional ability required to mitigate core damage or radioactive release" should be included within the scope of the rule. In attempting to evaluate which nonsafety systems should be included under this provision of the rule, the licensee found "significant fraction" a difficult criterion to use

in practice. Instead of using this criterion, the licensee included in the scope those non–safety-related systems considered to be "significant contributors to the mitigation strategy of the procedures." The licensee defined significant contributors as those "systems whose failure would affect the outcome of successful completion of the procedure or cause the user to transition to another procedure." The team concluded that the licensee's alternative criterion for identifying SSCs in EOPs is reasonable and meets the intent of the rule.

Table 1 is a summary of SSCs defined by the licensee as being within the scope of the rule at each site and any additional SSCs found by the team. The team found considerable variation even among similar plants in the numbers of SSCs defined for each plant and considerable variation in the numbers of SSCs identified as being within the scope of the rule. The numbers in this table are for general information and should not be used for judging the acceptability of the scoping activities at other sites.

Table 1. Structures, systems, and components under the scope of the Maintenance Rule[1]

	BWR/3	BWR/4	BWR/6	WEST 3 Loop	WEST 4 Loop	WEST 4 Loop	WEST 4 Loop	WEST 4 Loop	CE	B&W
Total number of SSCs	102	131	341	205	176	194	112	160	137	
Number of SSCs within scope	67 (66%)	86 (66%)	127[3] (37%)	115[4] (56%)	103[5] (59%)	100[6] (52%)	76[7] (68%)	110[8] (69%)	90[9] (66%)	
Number of structures (only) within scope[2]	n.a.[10]	6	16	15	6	17	23	32[8]	7[9]	
NRC-identified SSCs requiring reevaluation	None	None	1[3]	4[4]	1[5]	1[6]	3[7]	1[8]	15[9]	

1. The data on this table is based on a review of the licensee's documentation for rule implementation at the time of each NRC site visit and is subject to change.

2. Three licensees were still evaluating structures under scope at the time of the NRC site visits (i.e., These include the licensees for utility data in columns one, eight, and nine); therefore, the number of structures and the total number of SSCs under scope may change. All other licensees included structures under scope.

3. The licensee found 126 SSCs. The NRC recommended that control room annunciators be added.

4. The licensee found 111 SSCs. The NRC recommended that the site grounding system, the plant computer, heat tracing and freeze protection and the reactor coolant pump vibration monitor be added to the scope.

5. The licensee found 102 SSCs. The NRC recommended that the circulating water system be added to the scope.

6. The licensee found 82 out of 151 plant systems and 17 out of 43 plant structures within scope. The NRC recommended that the extraction steam system be added to the scope.

7. The licensee found 73 SSCs. The NRC recommended that the lighting protection, site grounding and cathodic protection be added to the scope.

8. The licensee found 109 SSCs under the scope of the rule; however, the NRC recommended that one additional structure, the shield wall around the startup transformers, be added to the scope.

9. The licensee found 75 SSCs. The NRC recommended that the circulating water, extraction steam, condenser air removal, screen wash water, gland steam, gland seal water, generator gas, turbine lube oil, and turbine generator seal oil systems be added to the scope. The licensee identified the reactor containment as the only structure that is under scope and being monitored. The licensee considers the remaining structures to be inherently reliable. The NRC recommended that 6 additional structures (i.e., the auxiliary building, control complex, emergency diesel generator building, intermediate building, NSSW intake structure, and the emergency feedwater tank enclosure) be added to the scope.

10. Utility continuing evaluation at the time of site visit.

Conclusion

The scoping at each site was thorough. The licensees correctly classified most safety-related SSCs (paragraph (b)(1) of the rule) and those non–safety-related SSCs whose failure could prevent safety-related SSCs from fulfilling their safety-related function (paragraph (b)(2)(ii) of the rule). The team believes the licensees should have included several other non–safety-related SSCs within the scope of rule: those relied upon to mitigate accidents or transients, those used in EOPs (paragraph (b)(2)(i) of the rule), and those whose failure could cause a reactor scram or actuation of a safety-related system (paragraph (b)(2)(iii) of the rule). The number of SSCs in this category at each site follows: 0 SSCs at 2 sites, 1 SSC at each of 4 sites, 3 SSCs at 1 site, 4 SSCs at 1 site, and 15 SSCs at 1 site.

Recommendations

1. Do not use the following reasons for excluding SSCs from the scope of the rule because they do not address the criteria in paragraph (b) of the rule.

   - The SSC is very reliable, inherently reliable, or has never failed at this site.

   - Redundant trains will prevent the system from every completely failing.

   - Operator actions will prevent the failure of the system from causing a scram.

   - The failure of the system will not directly cause a scram.

2. Review scoping determinations to ensure that no SSCs were excluded from the scope of the rule without adequate justification.

2.3   Risk Determination

The rule requires that goals be established commensurate with safety. To implement the rule in accordance with NUMARC 93-01, the licensee must do a risk (or safety) determination for all SSCs within the scope of the rule. This risk determination would then be considered when setting goals and monitoring under paragraph (a)(1) of the rule and when establishing performance criteria under paragraph (a)(2). The risk determination method recommended in NUMARC 93-01 involves the use of an expert panel using the Delphi method of NUREG/CR-5424, supplemented by probabilistic risk (or safety) assessment (PRA) or IPE insights, to find risk-significant SSCs. These PRA or IPE insights can include risk reduction worth (RRW), risk achievement worth (RAW), and core damage frequency contribution (CDF).

Findings

The team found that all licensees used an expert panel (or a working group) to determine risk significance. These expert panels considered PRA or IPE insights using the methods described in NUMARC 93-01 although with variation. Not all licensees took RRW, RAW, and CDF into consideration. One licensee considered only CDF and not RRW or RAW. Another licensee considered CDF and RAW but not RRW. Several licensees considered the Fussell-Vesely (F/V) importance measure in addition to CDF, RAW, and RRW.

The team found that licensees' PRA experts were very knowledgeable and were aware of the limitations of the use of PRA insights. One of these limitations is that all risk-important systems are not necessarily modeled in a PRA. Improvements can also be made in databases, success criteria (which affect accident sequence emphasis), and human reliability analyses. The team found an expert panel was necessary to compensate for the limitations and assumptions inherent in a PRA and to add a perspective of experience to the risk determination process. The team also found that, although CDF, RRW, RAW, and F/V all gave useful insights, none was indispensable as long as the results were reviewed and evaluated by a qualified expert panel.

The team interviewed members of the expert panel at each site and found that the panel members were knowledgeable and experienced and met the standards recommended in NUMARC 93-01. The expert panels made risk-significant determinations as recommended in NUMARC 93-01. Those at several of the sites participated in other maintenance activities such as scoping, establishing performance criteria and goals, determining when SSCs should be moved from paragraph (a)(1) to (a)(2) and from paragraph (a)(2) to (a)(1), reviewing corrective actions, and doing the periodic evaluation required by paragraph (a)(3) of the rule. However, some licensees considered the expert panel to be a temporary organization and the performance of the risk determinations to be a one-time activity. Other licensees, especially those who have decided to have their expert panel participate in other rule activities, have made the expert panel a permanent part of their rule program. The team believes that those licensees who do not have a permanent expert panel will need to reconstitute the expert panel in the future to revise these risk determinations as the plant is modified or when the PRA is revised.

Six of the nine licensees plan to update their PRA using plant-specific reliability and availability data to maintain a *living* PRA. The intervals for these updates are tentative and vary from one year to three years, or are undetermined. The remaining three licensees do not plan to update their PRAs routinely to incorporate plant-specific reliability and availability data. They did, however, state that they would evaluate the effect of any major plant modifications on PRA models, results, and conclusions and update them if necessary.

The staff at the last few sites visited were more aware of the value of time-structured PRA-determined core damage frequency profiles. These licensees had very strong PRA expertise and benefitted from the lessons learned from the site visits. The team believes that all licensees should share information and ideas among themselves.

Table 2 summarizes the number of SSCs that were determined to be risk significant, and the methods used at each site.[5]

---

[5] The team found considerable variation, even for reactors of similar type, in the numbers of SSCs defined for each plant and a considerable variation in the numbers of SSCs determined to be risk significant. The numbers in this table are for general information and should not be used for judging the acceptability of maintenance rule activities at other sites.

Table 2. Risk-significant structures, systems, and components and risk determination methods[1]

	BWR/3	BWR/4	BWR/6	WEST 3 Loop	WEST 4 Loop	WEST 4 Loop	WEST 4 Loop	WEST 4 Loop	CE	B&W
Number of SSCs within scope	67	86	127	115	103	100	76		110	90
Number of risk-significant systems	25 (37%)	41 (48%)	24 (19%)	44 (38%)	41 (40%)	23 (23%)	22 (29%)		27 (25%)	17 (19%)
Risk-significant structures	Utility evaluating	All structures are inherently reliable	All structures are inherently reliable	Reactor containment	Reactor containment; all other structures inherently reliable	Reactor containment; utility evaluating	Reactor containment		28	Reactor containment
Living PRA (yes/no) and update frequency	Yes. Update frequency undetermined	Yes, 3-year update frequency	No	No	Yes, Update frequency every refueling cycle	Yes, 3-year update frequency	Yes, Update frequency undetermined		No	Yes, Update frequency every refueling cycle
Risk determination methods[2]	EP Delphi process, IMs (RRW, RAW, F/V,)	EP Process IPE RISKMAN software, IMs (RAW, RRW, CDF contrib.)	EP Delphi process, IMs (RAW, RRW, CDF contrib.)	EP Process, IMs (RAW, CDF, contrib.)	EP Delphi process, IMs (RAW, RRW)	EP Delphi process, PRT software, IMs (RAW, CDF contrib.)	EP Process, PRT software, IMs (RAW, RRW, F/V)		Working group, IMs (RAW, CDF contrib.)	EP Delphi process, IMs (RAW, RRW)
Online maintenance risk evaluation methods[2]	CAFTA, RMQS and EOOS software, safety monitor	Configuration equipment OOS matrix	Configuration matrix, RAW ranking	PRa configuration matrix	Online maintenance PRA procedure	ORAM matrix	RMQS and OSPRE software		PRA procedure for taking SSCs OOS	PSAM risk monitor software

1. The data in this table is based on a review of the licensee's documentation for rule implementation at the time of each NRC site visit and is subject to change.

2. Acronyms for risk determination and evaluation: IM (importance measure), ORAM (outage risk assessment and evaluation: IM (importance measure), ORAM (outage risk assessment measure), ORAM (outage risk assessment and management), CAFTA (computer-aided fault tree analysis), PSAM (probabilistic safety assessment monitor), PRT (probabilistic risk tree), RMQS (risk management query system), EOOS (equipment out of service), OOS (out of service), OSPRE (operational safety predictor), F/V (Fussell-Vesely importance), RAW (risk achievement worth), RRW (risk reduction worth), CDF (core damage frequency; in accord with NUMARC 93-01, the utility defined cut sets that account for 90 percent of the overall CDF contribution), and EP (expert panel).

Conclusions

1. The methods used to establish risk significance met the intent of the rule and the guidance in NUMARC 93-01.

2. The risk determination process at each site gave reasonable results.

3. The use of an expert panel to consider PRA or IPE insights is an appropriate and practical method of determining risk significance.

4. The expert panel members at each site were knowledgeable and experienced and met the standards recommended in NUMARC 93-01.

5. The participation of the expert panel in other rule activities at certain sites is a strength.

Recommendations

1. Use the process described in NUMARC 93-01 which makes use of an expert panel for making risk determinations.

2. Have an expert panel evaluate CDF, RRW, RAW, F/V, and any other methods to compensate for their limitations.

3. Consider using other calculational methods such as Fussell-Vesely, Birnbaum, and others in addition to CDF, RHR, and RAW.

4. Establish the expert panel as a permanent part of the licensee's organization.

5. Consider using the expert panel to assist in making decisions on other aspects of implementing the rule such as scoping, the establishment of performance criteria and goals, the determination of when SSCs should be moved from paragraph (a)(1) to (a)(2) and from paragraph (a)(2) to (a)(1), the review of corrective actions, and the

performance of the periodic evaluation required by paragraph (a)(3) of the rule.

6. Reevaluate risk significance determinations whenever the plant design is modified, the PRA is updated, or new insights become available from configuration management reviews.

2.4 Goal Setting, Monitoring, and Preventive Maintenance

The team reviewed program documents and records at each site to evaluate the process established to set goals and monitor under paragraph (a)(1) of the rule and to verify that preventive maintenance was effective under paragraph (a)(2) of the rule. The team also discussed the program with plant staff. At each site, the team selected a sample of systems that were categorized under paragraphs (a)(1) and (a)(2) for further review.

2.4.1 Categorizing SSCs in Paragraph (a)(1) or (a)(2)

Regulatory Guide 1.160 and NUMARC 93-01 state that SSCs be subject to goal setting and monitoring under paragraph (a)(1) of the rule whenever performance criteria are exceeded or repetitive maintenance-preventible functional failures (MPFFs) occur.

Findings

The team reviewed a sample of systems at each site that were categorized under paragraph (a)(2) of the rule to determine if they should have been categorized under paragraph (a)(1), where they would have been subject to goal setting and monitoring. At eight of the nine sites, the team did not find any SSCs that should have been categorized under paragraph (a)(1) of the rule. At one site where the licensee had not categorized the salt service water system (SSW), the team recommended that it be categorized as a paragraph (a)(1) system.

Very few SSCs were categorized under paragraph (a)(1) at any of the sites. Table 3 presents the numbers of SSCs in each category at each site.[6] Licensees are still in the early stages of implementing the rule and thus may assign more SSCs to the paragraph (a)(1) category in the future. However, licensees may be reluctant to place SSCs in the paragraph (a)(1) category because having numerous SSCs in the paragraph (a)(1) category might indicate to the NRC or licensee management that their preventive maintenance program is not effective.

---

[6] The team found considerable variation, even for reactors of similar type, in the numbers of SSCs defined for each plant and in the numbers of SSCs placed in the (a)(1) category. The numbers in this table are for general information and should not be used for judging the acceptability of the maintenance rule activities at other sites.

Table 3. Structures, systems, and components categorized under paragraphs (a)(1) and (a)(2)[1]

Paragraph	BWR/3	BWR/4	BWR/6	WEST 3 Loop	WEST 4 Loop	WEST 4 Loop	WEST 4 Loop	CE	B&W
(a)(1)	1[2]	1[3]	4[4]	6[5]	3[6]	1[7]	5[8]	5[9]	3[10]
(a)(2)	66	85	123	109	100	99	71	105	87

1. The data on this table is based on a review of licensee's documentation for rule implementation at the time of each site visit and is subject to change.

2. At the time of the NRC site visit, the licensee was considering adding the salt service water (SSW) system to paragraph (a)(1).

3. The only SSC categorized under paragraph (a)(1) was the post-accident sampling system; however, the utility was still evaluating the performance of SSCs to determine if other SSCs should be placed under the paragraph (a)(1) category.

4. The SSCs categorized under paragraph (a)(1) include the residual heat removal system, the service water system, containment integrity and the ESF switch gear room coolers.

5. The SSCs categorized under paragraph (a)(1) include the turbine-driven auxiliary feedwater pump, the LK-16 circuit breaker, the BIF butterfly valves, the "B" emergency diesel generator, the reactor cavity seals, and the "A" heater drain pump motor.

6. The SSCs categorized under paragraph (a)(1) include the auxiliary feedwater system, the emergency diesel generators, and the solid state protection system (i.e., 7300 process support system).

7. The reactor coolant system was the only system categorized under paragraph (a)(1) of the rule.

8. The SSCs categorized under paragraph (a)(1) include the control rod drive system circuit cards, 4 kv circuit breakers, 125 vdc circuit breakers, steam generator power-operated relief valves (PORVs) and pressurizer PORVs.

9. The SSCs categorized under paragraph (a)(1) include the containment control air system, the condensate system, the heater drain system, the service water system, and the circulating water system.

10. The SSCs categorized under paragraph (a)(1) include the emergency diesel generators, the instrument air system and the demineralized water system.

The methods described in NUMARC 93-01 assume that the performance or condition of most SSCs at most sites is adequately controlled through preventive maintenance and therefore should, at least initially, be placed in category (a)(2), where they are monitored against performance criteria. Only those SSCs that exceed their performance criteria, or experienced repetitive maintenance preventible functional failures are placed in the paragraph (a)(1) category, where they are monitored against goals. Therefore, using the NUMARC 93-01 guidance, category (a)(1) could be used as a tool to focus attention on those SSCs that need to be monitored more closely and does not indicate maintenance program effectiveness. While using paragraph (a)(1) in this way is an acceptable approach for implementing the rule, the team believes that an approach that places most SSCs in the paragraph (a)(1) category and places in category (a)(2) only a few SSCs whose performance or condition is being effectively controlled through the performance of effective preventive maintenance, would also meet the intent of the rule.

The team assured the licensees' representatives that the NRC staff would not consider the placement of SSCs in the paragraph (a)(1) category as an indicator of a poor maintenance program nor would it be used in determining the grade in the maintenance area of the Systematic Assessment of Licensee Performance. The team also cautioned licensee managers that they should not view the number of SSCs in the paragraph (a)(1) category as an indicator of performance, since it might inhibit their staff members from placing an SSC under paragraph (a)(1) when a performance criterion was exceeded or a repetitive maintenance preventible functional failure had occurred. If a licensee is not certain whether or not an SSC should be categorized in paragraph (a)(1) or (a)(2), the conservative approach is to place the SSC in the paragraph

(a)(1). Failure to place the SSC under paragraph (a)(1) when preventive maintenance has been shown to be ineffective would be a violation of the rule.

Several licensees' procedures did not clearly explain the intended use of paragraph (a)(1) of the rule. The team recommended to these licensees that they review, and if necessary revise, their procedures to clarify the intent of paragraph (a)(1) of the rule and to ensure that SSCs are placed in that category whenever adequate preventive maintenance can no longer be demonstrated.

Conclusion

The process and procedures for categorizing SSCs in paragraph (a)(1) or (a)(2) of the rule were reasonable. However, some licensees were reluctant to place SSCs in the (a)(1) category.

Recommendations

1. Review, and if necessary, revise procedures to clarify the criteria for determining when goal setting under paragraph (a)(1) is required and to emphasize that the conservative approach to implementing the rule would be to categorize an SSC under paragraph (a)(1) whenever there was any doubt if the performance criteria had been met or a repetitive maintenance preventible functional failure had occurred.

2. The number of SSCs in the paragraph (a)(1) category should not be used as an indicator of a poor maintenance program.

2.4.2 Corrective Actions

Paragraph (a)(1) of the rule states that appropriate corrective action shall be taken when the performance or condition of an SSC does not meet established goals.

Findings

Many licensees have assigned the task of determining the root cause and developing corrective action to the responsible system engineer at each site. At some sites, the licensee's expert panel participates in this process.

At one site, the team reviewed corrective actions for six paragraph (a)(1) systems and found certain corrective actions were ineffective. For example, the turbine-driven auxiliary feedwater pump experienced an over-speed trip that was attributed to a faulty logic card, which was replaced. Three months later, two more over-speed trips occurred: the first was again attributed to a faulty logic card, which was also replaced, and the second was attributed to a faulty seal on the servo unit. Two months later, a fourth over speed trip occurred, which was again attributed to a faulty logic card. The team concluded that a more thorough evaluation of the first failure might have enabled the licensee to find the root cause of the logic card failure and avoid the subsequent failures. To enhance its corrective action process, the licensee began requiring that all proposed corrective actions be reviewed and approved by the expert panel. The corrective actions will also include a review of the corrective, predictive, and preventive maintenance activities. The licensee's improved process for establishing corrective actions is very rigorous and methodical.

The team reviewed the actions taken by one licensee to resolve problems with its post-accident sampling system (PASS). This system had been very unreliable in the past and was found to be inoperable when needed during the two most recent emergency drills. The team reviewed the corrective actions taken for PASS since the licensee began implementing the rule and noted that the system engineer had thoroughly reviewed past performance and industry operating experience to determine appropriate corrective actions. These corrective actions included unclogging lines, replacing leaking valves, revising procedures, and training PASS system operators. Long-term actions include the establishment of goals and a monitoring program. The team also interviewed the system engineer and found her to be very knowledgeable of the PASS system and the rule. The team believes that actions to implement the rule caused licensees to focus more attention to establishing appropriate corrective actions for PASS and other non–safety-related systems.

The team reviewed the corrective actions taken by another licensee to repair a faulty pressurizer pressure transmitter. The system engineer had performed extensive reviews, considered industry-wide operating experience, and trended the performance of similar transmitters to determine the cause of the faulty transmitter. The corrective actions established in this process corrected the problem.

The licensees for most of the other sites visited by the team had established effective corrective action programs. At these sites, the system engineer was generally assigned responsibility for determining corrective actions.

Conclusions

Licensees established effective corrective action programs. Some of those programs improved significantly after the licensees acted to the implement the rule, while others appear to have been effective before the rule. Although most licensees assigned the system engineers primary responsibility for establishing corrective action, a few assigned that responsibility to an expert panel. Both approaches produced acceptable results.

Recommendation

None.

2.4.3 Safety Consideration in Goal Setting

Paragraph (a)(1) of the rule requires that safety (risk) be considered when establishing goals and monitoring.

Findings

Each licensee performed risk determinations for all SSCs within the scope of the rule at its site. These risk determinations formed the basis for considering safety when setting goals under paragraph (a)(1) and performance criteria under (a)(2) of the rule. All licensees used the results of these initial risk determinations to decide if goals and performance criteria would be set at the system level or the plant level. System or train level goals or performance criteria were set for risk-significant SSCs and for non–risk-significant SSCs that were used in standby service. Plant level goals were set for the remaining non–risk-significant SSCs. Therefore, all licensees had taken safety into consideration through the process of determining whether to set system- or plant-level goals.

Some of the licensees had taken safety into consideration in other ways in addition to the steps described above. For example, the expert panel at one site received the information from the risk determination process to use in establishing goals for those SSCs assigned to categories under paragraph (a)(1) of the rule and in establishing performance criteria under paragraph (a)(2) of the rule. The expert panel decided to establish functional failures and hours unavailable as goals for most systems. To determine what would be appropriate unavailability values, the expert panel reviewed historical system performance for the risk-significant and standby systems within the scope of the rule. Since precise system unavailability data was not readily retrievable from plant records, the expert panel relied on their collective judgment to estimate system unavailability data from the monthly operating reports. To validate this unavailability data, the licensee recalculated the PRA using the new unavailability values and confirmed that the results were consistent with the original PRA calculation. This is another example of how safety has been considered when setting goals and performance criteria.

At another site, the licensee selected goals and performance criteria based on unavailability or reliability data assumed in the licensee's PRA. The team reviewed the goals set for the emergency diesel generator, the demineralized water system, and the instrument air system, and verified that information from the PRA had been considered. The team concluded that this was an acceptable method of taking safety into consideration when setting goals under paragraph (a)(1) of the rule.

Conclusions

1. All licensees had considered safety as part of the process of determining if an SSC should be categorized as risk-significant or non-risk-significant.

2. Certain licensees considered safety by using the same values for system reliability and availability for goals as were assumed in the PRA. Other licensees used the PRA to check the validity of availability goals.

Recommendation

Ensure that procedures and processes adequately address safety when setting goals.

2.4.4 Industry Operating Experience in Goal Setting

Paragraph (a)(1) of the rule requires that industry-wide operating experience (OE) be taken into account, where practical, when establishing goals.

Findings

The degree to which OE was considered when setting goals varied considerably among the sites. At one site, the team verified that OE information from the Institute for Nuclear Power Operations, and industry codes and standards had been considered when setting goals for some SSCs under paragraph (a)(1) of the rule but not for others. The team reviewed the licensee's procedures and noted that they only required that relevant industry NPRDS operating experience be reviewed and considered as part of the goal setting process. The licensee agreed to revise their procedure to clarify that their review should not be limited to NPRDS data.

At another site, the team reviewed the goals set for several systems and verified, through discussion with the licensee's system engineers, that OE had been considered when setting goals. However, the team reviewed the licensee's procedures and found no explicit requirement to consider OE as part of the goal setting process. The licensee's representative stated that consideration of OE was part of the root cause analysis process that would be required whenever goals were established. However, the representative agreed to add the requirement to consider all relevant information from their OE program as part of their goal setting procedure.

Several licensees had considered OE when establishing goals for some of their paragraph (a)(1) systems but not for others. These licensees agreed to revise existing procedures or issue a new procedure to clarify that OE

must be taken into account whenever goals are established. At several other sites, the licensees had considered OE when establishing goals for all their paragraph (a)(1) systems and their procedures were adequate.

At some sites, the OE database was available to each engineer on his or her computer terminal, which could be used at any time to research for information on system or component problems. The team believes that such an online capability would facilitate the process of considering industry operating experience when setting goals. At other sites, the process for obtaining information from the OE database required individual engineers to complete request forms, which were sent to the OE database manager who did the actual search. This process is more cumbersome than having an online capability and could inhibit the use of OE for goal setting.

During the site visits, the team did not mention its concern that licensees had not established a systematic and consistent method of collecting and using SSC reliability and availability data from other licensees when setting goals. The team came to this conclusion during internal NRC meetings to discuss the results of the nine site visits. The team understands that goals for some SSCs are based on PRA reliability and availability data which, to some extent, may have been developed from generic industry data. Therefore, industry-wide reliability and availability data may have been considered when setting goals for a limited number of SSCs. However, after considering this issue in the light of the results of all nine pilot inspections, the team concluded that a more direct use of industry-wide reliability and availability data should be considered by licensees. Existing OE programs at most sites focus on anecdotal data such as descriptions of specific equipment failures and do not generally give reliability and availability data. Therefore, the team recommends that licensees consider expanding their programs for collecting OE information to ensure that industry-wide SSC reliability and availability data is collected and considered in setting goals in a systematic and consistent manner.

Conclusions

1. Most licensees had considered OE in varying degrees when setting goals.

2. Many licensees' procedures do not have adequate guidance for ensuring that OE is considered, where practical, when establishing goals.

3. Those responsible for establishing goals at some sites had easy access to the OE database; at other sites the access was limited or cumbersome and could inhibit the use of the database.

4. Licensees had not established a systematic and consistent method of collecting and using SSC reliability and availability data from other licensees when setting goals.

Recommendations

1. Review procedures to ensure that the guidance is adequate for considering OE when establishing goals.

2. Ensure that OE data is readily accessible for plant staff to use when setting goals.

3. Consider expanding programs for collecting OE information to ensure that industry-wide SSC reliability and availability data is collected and considered systematically and consistently when setting goals

2.4.5 Monitoring and Trending of Systems and Components

In the statements of consideration for the rule, the Commission states that where failures are likely to cause loss of an intended function, monitoring under paragraph (a)(1) should be predictive, giving early warning of degradation. NUMARC 93-01 gives guidance for using predictive maintenance, inspection, testing, and performance trending for monitoring performance or condition under paragraph (a)(2) of the rule.

Findings

The team reviewed the monitoring and trending for selected systems and components at each site and found the following issues.

Coordination of trending and goals. The team found great variance among the licensees in the quality and quantity of trending that was being performed. One licensee was doing very little trending of SSCs performance or condition. Two other licensees had established trending programs that were well integrated into their rule programs. The remaining licensees had trending programs that were not well integrated into their rule programs. Many of these trending programs generated equipment performance data that would be very useful when establishing goals and performance criteria under the rule; however, in many cases, licensees did not consider this data when selecting goals and performance criteria and establishing a monitoring program under the rule. The team believes that goal setting and trending activities should be coordinated and integrated as much as possible so that the improvements in performance can be monitored against established goals. Goals should use existing trending activities where appropriate, and licensees should consider establishing new monitoring and trending activities that directly address the problem whenever new goals are established.

Trending for all goals not required. Although trending of all goals should always be considered, it may not be practical. For example, one licensee had established a predictive maintenance program that included periodic monitoring, diagnosis, and trending of system performance and condition so that needed maintenance could be performed before failure. Predictive maintenance actions included the use of diagnostic test equipment to perform vibration analysis, thermography, flow measurements, and ultrasonic measurement of pipe wall thickness. The team reviewed the documentation for five paragraph (a)(1) systems at that site and noted that comprehensive trending was being performed for at least two of these systems. The team concluded that trending was not necessarily required for the other three systems. The team believes that trending should be used where practical and appropriate, but not necessarily for all SSCs under the scope of the rule. Therefore, the level of trending being performed was reasonable.

Monitoring of redundant trains. The NRC staff endorsed the position in NUMARC 93-01 that systems with redundant trains must be monitored at the train level to ensure that the good performance of one train does not mask the poor performance of the redundant train in a system. This train-level monitoring would be required for all those SSCs that were determined to be risk-significant and all non–risk-significant SSCs that were used in standby service. Monitoring at the train level would not be required for normally operating non–risk-significant SSCs, which can be monitored using plant level goals.

The team noted several instances where redundant trains were not being monitored at the train level.

1. The fuel oil system for the emergency diesel generators at one site consisted of two redundant fuel oil pump trains, either of which can be used to transfer fuel oil from the large storage tanks to the diesel day tanks. The licensee had not established individual performance criteria and monitoring for each fuel oil pump train under the rule. The team was concerned that the redundant pump could degrade significantly without being detected because one pump could supply the necessary volume of fuel during routine surveillance tests. The high reliability of one pump could mask the unreliability of the redundant pump.

2. One licensee used multiple redundant air compressors to supply site compressed air and to act as an alternate supply for the instrument air system. The licensee's representatives stated that one of the reasons for the many compressors was that some of them were very unreliable. However, the licensee was monitoring this system using plant-level performance criteria instead of train-level performance criteria. The licensee's representative stated that they considered the extra compressors to be installed spares rather than redundant loops and therefore did not require train level goals for each compressor. The team disagreed with this position, noting that the intent of the rule as explained in NUMARC 93-01 is to monitor any system with redundant trains at the train level. The intent of the rule is also to monitor all standby systems at the system level rather than at the plant level. In this case, the air compressors constitute a system with multiple redundant compressor trains, some of which are used in standby service. Therefore, this system should not be monitored at the plant level both because it contains redundant trains and because some of those trains are used in standby service. Rather, each train, including the motor, compressor, and valves, should be monitored individually.

At another site, the containment spray system, which contains redundant trains, was being monitored at the system level rather than the train level, which could allow unreliable components to go undetected.

Trending of zero failures. At many of the sites visited, licensees had established zero MPFFs or 100-percent reliability as a goal or performance criterion for many of the SSCs under the scope of the rule. The rule intends that licensees be afforded maximum flexibility in establishing goals and performance criteria. However, the rule also intends that where failures are likely to cause loss of an intended function, monitoring should be predictive, giving early warning of degradation. The team was concerned that it would be difficult to use trending to help predict or anticipate failures when failure data is the only information being monitored.

Conclusions

Coordination of trending and goals. Most licensees had established trending programs. Trending was not required for all SSCs under the rule although it should be considered. The trending being performed by most licensees was not well coordinated and integrated with the goals and performance criteria established under the rule.

Monitoring of standby systems or systems with redundant trains. Certain non–risk-significant systems used in standby service were being monitored at the plant level rather than at the system or train level as required.

Trending of zero failures. Reliance on the use of zero MPFFs or 100-percent reliability as a goal or performance criterion may preclude predictive trending.

Recommendations

Coordination of trending and goals. Coordinate and integrate goals and

performance criteria with equipment trending wherever possible.

Monitoring of standby systems or systems with redundant trains. Ensure that one train does not mask the poor performance of a redundant or standby train. Accomplish this task by monitoring at the following levels.

1. Monitor single train risk-significant systems at the system level.

2. Monitor multiple train risk-significant systems at the train level.

3. Monitor single train non–risk-significant systems used in standby at the system level.

4. Monitor multiple train non–risk-significant systems that are used in standby service at the train level.

5. Monitor normally operating, non–risk-significant systems at the plant level.

Trending of zero failures. Where reliance on the use of zero MPFFs or 100-percent reliability as a goal or performance criterion may preclude predictive trending, consider establishing additional goals and performance criteria that can be trended.

2.4.6 Monitoring and Trending of Structures

The rule requires that the performance or condition of structures be monitored in a manner sufficient to give reasonable assurance that those structures are capable of fulfilling their intended function. The statements of consideration for the rule states "[w]here failures are likely to cause loss of an intended function, monitoring should be predictive in nature, providing early warning of degradation." NUMARC 93-01, paragraph 9.4.2.4 lists examples of structural monitoring activities including nondestructive examination, visual inspection, vibration analysis, and measurement of deflection.

Findings

The team reviewed the monitoring and trending of structures at the nine pilot sites and found that most licensees considered monitoring of structures under the rule to be a low priority. Some licensees had not established goals or performance criteria for monitoring most structures at their sites. Many licensees considered most structures to be inherently reliable. Some licensees believed that as long as a structure such as a building did not fall down and damage the equipment inside, the structure itself need not be monitored. At some sites, the onsite personnel were not aware of existing preventive maintenance and monitoring activities that were being performed on these structures by offsite structural or civil engineering groups.

One licensee took the position that their structures had performed acceptably for the past 20 years, not causing a loss of function of the systems contained in or supported by the these structures, and were not expected to begin a more rapid degradation from aging in the future. Therefore, believing these structures "very reliable," they found it unnecessary to establish goals or performance criteria to monitor them. However, they also stated that inspection and maintenance is necessary to ensure degradation of these structures does not cause a loss of function. The structures are monitored during the normal operator rounds, management walkaround inspections, and inspection by other plant departments in the course of normal work activities. Deficiency cards and maintenance work orders are generated when conditions adverse to quality are found. The team believes that the existence of these longstanding monitoring activities contradicts the licensee's position that no monitoring is

needed. The team believes that the licensee should establish performance criteria and goals under the rule which take credit for the existing monitoring activities and build upon them.

One licensee had established performance criteria for most structures under the scope of the rule. However, the performance criteria for many of these structures was that the structure would not degrade to the point where it caused a loss of function of systems contained in the structure or supported by it. For example, the roof of a building that was leaking would meet the performance criteria until the water leaking into the building caused the system inside the building to fail. However, such performance criteria are not acceptable because they are not predictive and do not give early warning of degradation. The team believes that a more appropriate performance criterion would have been "no water leaks."

Another licensee had determined that all structures within scope of the rule, except the primary containment, were inherently reliable and therefore did not require goal setting under paragraph (a)(1) of the rule or monitoring against performance criteria under paragraph (a)(2) of the rule. The licensee's representative stated that these structures are routinely examined by plant personnel during their walkdown inspections of the plant. They believe that this monitoring activity is sufficient to verify that preventive maintenance is adequate. The team believes that although condition-monitoring is an appropriate method of monitoring structures, the lack of specific criteria to monitor against would make it difficult to detect degradation of these structures.

Conclusions

Most licensees considered the monitoring of structures under the rule to be a low priority. Some licensees incorrectly assumed that many of their structures are inherently reliable. The performance criteria for monitoring the performance or condition of some structures are not predictive and do not give early warning of degradation.

Certain structures such as the primary containment can be monitored by fulfilling established testing requirements such as those in 10 CFR Part 50, Appendix J. However, other structures such as reactor buildings, auxiliary buildings, and cooling towers may be more amenable to condition-monitoring. Some licensees are developing programs for monitoring structures that will include doing plant walkdown inspections and engineering evaluations to establish condition-monitoring criteria. This program should include the establishment of specific criteria for monitoring.

Recommendations

1. Reevaluate the monitoring of structures and determine, using the methods described in NUMARC 93-01, paragraph 9.3.2, to determine whether performance criteria or goals are needed to monitor the performance or condition of individual structures.

2. Review the existing structural monitoring activities and use them, with enhancements as necessary, as a basis for establishing a monitoring program under the rule.

3. Do not use *very reliable* or *inherently reliable* to describe structures that require preventive maintenance or monitoring.

4. Establish performance criteria or goals that are predictive and give early warning of failure.

5. Take credit for existing plant walkdown inspections or other structural inspection activities under the rule.

2.4.7 Functional Failures

The statements of consideration for the rule state that where one or more *maintenance preventible failures* occur on SSCs under paragraph (a)(2) of the rule, the effectiveness of preventive maintenance is no longer demonstrated, and the SSC must then be treated under paragraph (a)(1) of the rule. This term was changed in NUMARC 93-01 to *maintenance preventible functional failures* to emphasize that only a failure in which the item actually failed the function should be counted as a failure that would require goal setting and monitoring under paragraph (a)(1).

Findings

Two licensees focused on *functional failures* rather than just *maintenance preventible functional failures* as described in NUMARC 93-01. These licensees did so because it was easier to process both types of failures (maintenance-related or not) in the same manner because they would not know if the failure was maintenance-preventible until after the root cause evaluation had been performed. The team reviewed this approach and noted that all functional failures would be evaluated and dispositioned in the same manner as maintenance-preventible functional failures.

Conclusion

The use of *functional failures* in stead of *maintenance-preventible functional failures* is acceptable. Approaches other than those described in NUMARC 93-01 are acceptable if the licensee can demonstrate that the alternative gives the same level of assurance that the requirements of the rule will be satisfied.

Recommendation

None

2.5 Periodic Evaluations

Paragraph (a)(3) of the rule requires that performance- and condition-monitoring activities and associated goals and preventive maintenance activities be evaluated at least every refueling cycle if the interval between evaluations does not exceed 24 months. The evaluation is required to consider, where practical, industry-wide operating experience.

At each site, the team reviewed the licensee's plans and procedures for performing the evaluations.

Findings

One licensee planned to do the periodic evaluation each year to coincide with other existing maintenance evaluation activities. Since the licensee's refueling cycle is longer than one year, performing the periodic evaluation annually would result in it being evaluated more frequently than required by the rule. At the remaining sites, the licensees planned to perform the evaluation once each refueling cycle.

At one of the two unit sites where a periodic evaluation is planned once each refueling cycle, the licensee plans to evaluate both units at the site at the same time. The licensee will periodically evaluate each unit at a different time during each unit's refueling cycle because the refueling outages for each unit at that site are not scheduled at the same time.

Another licensee plans to perform most of the required evaluations on a continuing schedule throughout the refueling cycle. System engineers would continuously assess the appropriateness of the performance criteria for their assigned systems and update performance data each month. In addition, a bimonthly system status report would be issued which would include an assessment of actual performance against the performance criteria for all systems within the scope of the rule, an evaluation of maintenance effectiveness, an evaluation of the balance of unavailability and reliability, and an evaluation of the continued applicability of the established performance criteria. These continuing activities would be followed by a high-level periodic evaluation to verify all rule requirements have been met. The evaluation would include a sampling of actual implementation. These high-level periodic evaluations would be performed each refueling cycle, not to exceed 24 months between evaluations.

Two licensees had already done periodic evaluations even though the first evaluation would not be due until after the rule takes effect on July 10, 1996. The licensees performed these evaluations to gain experience and to evaluate their progress in implementing the other requirements of the rule. The team reviewed these preliminary evaluations and found that both generally met the requirements of the rule.

Conclusions

1. Performing the periodic evaluation each year meets the intent of the rule if the refueling cycle is longer than one year.

2. The periodic evaluation does not have to be performed at any particular time during the refueling cycle as long it is performed at least one time during the cycle, and the interval between evaluations does not exceed 24 months. This requirement would permit the licensee for a multiple unit site to perform the periodic evaluations of all units at the same time even though the refueling cycles for the units are staggered.

3. The requirement for performing the periodic evaluation can be satisfied by using continuing evaluations, with a higher level summary evaluation performed at least once each refueling cycle.

4. The early periodic evaluations performed by two of the licensees were generally satisfactory.

5. The other seven licensees had reasonable preliminary plans and procedures for performing the periodic evaluation, although their implementation could not be reviewed at the time of the site visit.

Recommendations

None.

2.6   Balancing Unavailability and Reliability

Paragraph (a)(3) of the rule states that adjustments shall be made where necessary to ensure that the objective of preventing failures of structures, systems, and components through maintenance is appropriately balanced against the objective of minimizing the effect of monitoring or preventive maintenance on the availability of structures, systems, and components.

At each of the sites, the team reviewed the plans, processes, and procedures established for doing this activity.

Findings

The team found that although all licensees had established preliminary plans for accomplishing this activity, none of them had fully developed and implemented these plans. However, five of the licensees had procedures for accomplishing this activity, which were reviewed by the team.

Two licensees plan to continually balance unavailability and reliability as an integral part of monitoring against performance criteria under the rule. Performance history, preventive maintenance activities, and out-of-service time are considered when developing the performance criteria. The licensees believe that meeting these performance criteria will ensure that a satisfactory balance of reliability and unavailability has been achieved. The team found that guidance, acceptance criteria, or both needed to be added to the licensees' procedures.

At one site, unavailability and reliability were initially balanced by the licensee's expert panel. The licensee will maintain this balance by trending and evaluating SSCs rendered unavailable because of preventive maintenance and by making adjustments where necessary to achieve the appropriate balance. The licensee's Performance Monitoring Group will perform this activity as part of their semi-annual evaluation of maintenance program activities.

One licensee plans to accomplish this balancing by calculating the risk contribution associated with unavailability of the system caused by preventive maintenance activities and the risk contribution causing the reliability of the SSC. The licensee would then attempt to balance the contribution to risk from each source to ensure consistency with PRA or IPE evaluations.

Another licensee's PRA was used to determine values for unavailability and reliability which, if met, would ensure that certain threshold CDF values would not be exceeded. The performance criteria were established in accordance with these unavailability and reliability values. The licensee believes that meeting these performance criteria will ensure a reasonable balance of unavailability and reliability is attained.

Conclusions

Each of the nine licensees established reasonable preliminary plans for balancing unavailability and reliability, and the procedures developed by five of those licensees were reasonable. However, the team was unable to evaluate these activities because these plans had not been implemented at the time of the site visits.

Recommendation

Focus attention on establishing and implementing processes and procedures for balancing unavailability and reliability as required by paragraph (a)(3) of the rule.

2.7 Plant Safety Assessments for Taking Equipment out of Service

Paragraph (a)(3) of the rule states that, when performing monitoring and preventive maintenance activities, an assessment of the total plant equipment that is out of service should be considered to determine the overall effect on performance of safety functions.

NRC issued Temporary Instruction (TI) 2515/126, "Evaluation of On-Line Maintenance," to aid inspectors in evaluating the effect on safety of licensee procedures and practices for removing equipment from service for online maintenance. This instruction details the NRC's expectations regarding safety assessments to be performed before taking equipment out of service. The instruction recommends inspectors consider three factors when evaluating the overall risk of taking equipment out of service for online maintenance: (1) the probability of an initiating event such as a loss-of-coolant accident, turbine trip, or loss of offsite power; (2) the probability of being able to mitigate the event using core damage prevention; and (3) the

probability of being able to mitigate the consequences by preserving containment integrity.

Findings

One licensee elected to use a matrix approach, which involved listing preanalyzed configurations to supplement their existing procedural guidance for voluntary online maintenance. This list of preanalyzed configurations was developed using risk achievement worth (RAW) to rank configurations according to risk. The licensee's existing work planning processes established a preventive maintenance (PM) schedule, which allows work on only one division (train) of certain systems to help limit the risk of placing multiple trains of the same system out of service at one time.

One licensee makes use of a 12-week *rolling window* cycle to schedule preventive maintenance activities. At the time of the site visit, the licensee was developing a procedure to establish PRA-based priorities for taking certain combinations of systems out of service for maintenance at the same time.

One licensee's plant operations staff maintained an equipment out-of-service (EOOS) status board to assess the total plant equipment that is out of service. To control risk in the schedule for the upcoming maintenance activities, the licensee will use the plant IPE to evaluate the increases in core damage frequency (CDF) resulting from multiple equipment outages. The licensee will adjust the schedule to minimize significant spikes in the CDF envelope of the schedule. The licensee has also established administrative controls that forbid certain equipment out-of-service configurations under specified conditions. In addition, the licensee is considering the purchase of a real-time safety (risk) monitor.

One licensee plans to incorporate a matrix of risk-significant combinations of equipment into their 12-week work planning schedule. This matrix, which was developed from the licensee's PRA, lists combinations of equipment that would increase risk unacceptably if taken out of service at the same time. Procedures to implement this process had not been developed at the time of the site visit.

One licensee is planning to design a rolling windows (12-week) maintenance schedule with the matrix formulation of the Operational Safety Predictor (OSPRE) methodology to control the risk resulting multiple equipment outages. The licensee will adjust this schedule with the OSPRE matrix to minimize risk. In addition, the licensee drafted a policy statement that requires the establishment of an effective risk management program to determine the relative amount of risk involved in making systems or components unavailable and to evaluate the effect of multiple concurrent outages of systems important to safety. This policy statement addresses the need to (1) assess the confidence in the redundant equipment before taking equipment out of service for maintenance, (2) minimize challenges to that redundant equipment, and (3) establish contingency plans in case the redundant train is also rendered inoperable.

One licensee's online maintenance work philosophy is based on "ensuring plant safety, maximizing availability, and maintaining adequate reliability." The plant operations staff maintains a "plan of the day" that describes all work authorized to be performed on or around operating plant equipment during the 24-hour period. Specific procedures for establishing a preventive maintenance equipment out-of-service matrix were still being developed at the time of the site visit.

One licensee plans to evaluate the total effect on plant safety by using an IPE tool and hand calculations. This IPE tool will be used for evaluating the risk significance of online maintenance strategies as an interim approach until more sophisticated calculational aids are developed. The IPE tool gives risk evaluations for the 500 highest contributing core damage sequences, which account for about 93 percent of the total CDF. The licensee is considering a more accurate tool for the longer term, such as a risk monitor (near real-time), which would be used for preventive maintenance planning and other applications. The licensee is preparing the procedures to implement this requirement. In addition, this licensee plans to design a rolling windows preventive maintenance schedule with a matrix formulation methodology to control the risk resulting multiple equipment outages in the various schedular weeks. The licensee will adjust this schedule with the matrix to minimize risk.

One licensee implemented a program that generates risk profiles for planned maintenance activities each week. These profiles are discussed at a weekly scheduling meeting. Changes are made to the schedule to improve safety when undesirable equipment outage configuration changes are anticipated. The licensee also plans to establish a rolling window (12-week) maintenance schedule that will use PRA methodology to control the risk resulting from multiple equipment outages. They also plan to use the outage risk assessment and management (ORAM) tool to control shutdown risk associated with maintenance activities during outages.

One licensee is using probabilistic safety assessment monitor (PSAM) software to improve the reliability and safety of the plant by enhanced online risk monitoring of risk-significant systems included in the plant PRA. A weekly draft online maintenance schedule is produced on Thursday of each week and is analyzed using the risk monitor software before noon on Friday of the same week. On Monday of the following week, before equipment is taken out of service for maintenance, a final risk analysis is performed and any necessary schedular adjustments are made. The licensee also reduces the occurrence of initiating events by paying heightened attention to main feedwater system operability before performing maintenance on emergency feedwater pumps and by ensuring that no switchyard work is performed when an emergency diesel generator is out of service. The licensee analyzes plant conditions each day, including reviewing operational logs to ensure that redundant train equipment or support equipment is not degraded during the online maintenance. The licensee uses a software program called CAFTA (Computer Aided Fault Tree Analysis) to aid in controlling shutdown risk.

Conclusions

Certain licensees were using a variety of approaches for assessing the overall effect on the performance of safety functions of taking plant equipment out of service for monitoring or preventive maintenance, while others were planning to do so.

Some licensees plans and procedures for performing safety assessments appear to be well thought out and comprehensive although they had not been fully implemented. The plans at other sites were preliminary, and the procedures had not yet been developed or implemented.

Many licensees approached the problem by developing a matrix that defines which system combinations could be allowed out of service at the same time. Although the matrix is simple to use, it defines a limited number of combinations that may not address all

operational situations and may unnecessarily limit operational flexibility.

Several licensees are planning to use real time (or nearly real time) risk monitors that can calculate the risk changes associated with the planned maintenance activities. Although risk monitors can analyze a greater number of possible combinations of out of service systems, they may require specially trained personnel to operate them or to interpret the results.

Both the matrix and the risk monitor were reasonable ways of assessing the effect on plant safety when taking equipment out of service for monitoring or preventive maintenance. However, the team could not evaluate the effectiveness of either of these methods because they had not been fully implemented at the time of the site visits.

Many of the licensees had not established programs and procedures that prescribed giving consideration to initiating events, mitigating capability, and containment integrity before taking equipment out of service for monitoring or preventive maintenance.

Recommendations

1. Develop and implement processes and procedures for considering plant safety before taking equipment out of service for maintenance.

2. Include a program requirement to consider initiating events, mitigating capability, and containment integrity before taking equipment out of service for monitoring or preventive maintenance.

3. Pay increased attention to reducing initiating event frequencies, with particular emphasis on enhanced safety oversight of switchyard activities.

3 CONCLUSION

The team considered the observations made during these pilot site visits and concluded that the requirements of 10 CFR 50.65 can be met by using NUMARC 93-01 if the recommendations in this report are taken into consideration.

The team also concluded that the performance-based approach to implementing the rule is practical; the draft inspection procedure can be used to monitor the implementation of the rule; and the existing PRA tools and models, used in conjunction with an expert panel, are adequate for taking risk into consideration when implementing the rule.